Amateur Radio for the Antisocial

It's not all about the ragchew

By Allan Hall

Copyright © 2020 by Allan Hall
10 9 8 7 6 5 4 3 2 1

All rights reserved. No part of this publication may be reproduced, distributed, or transmitted in any form or by any means, including photocopying, recording, or other electronic or mechanical methods, without the prior written permission of the publisher, except in the case of brief quotations embodied in critical reviews and certain other noncommercial uses permitted by copyright law. For permission requests, write to the publisher, addressed "Attention: Permissions Coordinator," at the address below.

Allan Hall
1614 Woodland Lane
Huntsville, TX 77340
www.allans-stuff.com/books/

Although the author and publisher have made every effort to ensure that the information in this book was correct at press time, the author and publisher do not assume and hereby disclaim any liability to any party for any loss, damage, or disruption caused by errors or omissions, whether such errors or omissions result from negligence, accident, or any other cause.

Any trademarks, service marks, product names or named features are assumed to be the property of their respective owners, and are used only for reference. There is no implied endorsement if we use one of these terms.

All images/graphics/illustrations in this book are copyrighted works by Allan Hall, ALL RIGHTS RESERVED or in the public domain.

Acknowledgements:

As always I have to thank my wife, Sue Ann, because without her prodding and help, my books would either never get done or would be far less than they are.
I would also like to thank my readers. Your emails, letters and reviews let me know that my work is appreciated which drives my desire to create more and better books.

Table of Contents

1: Introduction 1

2: Things to work on 3

    2.1: Build a radio 5

    2.2: Repair vintage equipment 7

    2.3: Build and test antennas 9

    2.4: Bake a pi! 13

    2.5: Get an Arduino 15

    2.6: Help others with a blog or website 17

3: Things to do 19

    3.1: Collect QSL cards 21

    3.2: Chase awards 23

    3.3: Radio astronomy 25

    3.4: FT8 digital 27

    3.5: SDR dongles 31

    3.6: Pirate hunting, arrrr 33

    3.7: QRP 35

    3.8: Working satellites 37

    3.9: Moon bounce 39

    3.10: Meteor scatter 41

    3.11: Slow scan TV 43

    3.12: Amateur TV 45

    3.13: Kite antennas 47

    3.14: Weather Fax 49

3.15: QRSS 51
3.16: SOTA summit hiking 53
3.17: APRS 55
3.18: WSPR low power beaconing 59

4: Things to participate in 61
4.1: Contesting 63
4.2: Skywarn 69
4.3: Local club events 71

5: In closing 73

6: More information 75
6.1: Index 77
6.2: Glossary 78
6.3: Q Codes 89
6.3: ITU Phonetic Alphabet 91
6.4: Other books by the author 93
6.5: Notes 95

Introduction

I love technology in pretty much all its forms. From old school tubes to new SDRs, from computers to internet connected thermostats, it is all fascinating. Unfortunately I am not really into meetings, group outings, and long conversations about growing petunias. I am fairly antisocial.

This is often seen as a barrier to amateur radio as people seem to think it is all about two or more people "rag chewing", having those aforementioned long and seemingly pointless conversations. After all, what can you talk about that you could not do better, and more efficiently using email?

Even worse is the look on someone's face when they first hear a "net" and they turn to me and ask if there was actually a point to what just happened.

Some of you are already wondering, isn't amateur radio about communications? How can you enjoy a hobby that centers on communications without actually communicating?

The great news is that there is such a wonderful array of things in amateur radio that you can do with virtually no conversations, or at least no verbal conversations longer than a minute or so. You can have a long lasting and enjoyable hobby with amateur radio without any chewing at all.

Even better is that you can pick and choose what works for you, there is no one-size-fits-all in amateur radio. There also is no compendium as to what all there is to do in the hobby, it is continually growing and changing as new people and new technologies get involved.

What is presented here is simply something to whet your appetite and get you started down the road to having some fun. A small collection of things requiring little to no long conversations, no boring meetings, and no bumping elbows with thousands of other people at some kind of convention.

Yes, you will need to communicate, and that is just as much fun for me as it is for an old-timer who wants to yack your ear off for hours at a time with no real point to the conversation. The difference is that my idea of communication is to exchange the data that needs to be exchanged, say a pleasantry or two, and get on to the next task at hand.

If this sounds like you, then you are probably going to really enjoy this book and the hobby as a whole.

I should point what this book is not, and that is a tutorial for getting your license. I am assuming you either have a license or are prepared to get one already and are just making sure this hobby is for you.

So sit back, strap in, and get ready for a ride…

Things to work on

Build a radio

One of the oldest radio related aspects I can think of is actually making your own radio from a set of parts, either in a kit or just from some plans.

I remember doing this from a Radio Shack kit I got for Christmas probably back in the 1970s. The kit required no soldering, but instead used springs where you placed the pre-striped portions of the jumper wires to complete the circuits.

I can remember using the drain pipe of my bathroom sink as a ground and eventually using the cable TV wire as an antenna. Those were fun days and it was amazing what all you could hear. I was in south-east Texas and routinely listened to a radio station out of Chicago at night when the weather was good.

Radio Shack sold a large selection of these kits back in the day and I recall wanting just about all of them, and actually got a few of them. I was fortunate that my father was an engineer and enjoyed feeding my inner technology junkie.

Unfortunately, today we have no more Radio Shack per se, they are more mainstream and less electronic geek supplier. That's OK though because we now have the internet where you can get just about anything!

For younger readers, or parents thereof, I like the Haynes Retro Radio Kit as it resembles a completed actual radio and works pretty dang well and does not require soldering. Next up might be something like the Elenco AM/FM Radio Kit that does require soldering. This is the perfect kit to learn more about how a radio actually works while learning or honing your soldering skills.

Yes, those are little kits for listening, which anyone can do. Why did you get an amateur radio license if all you are going to do is listen? I put those in here for those of you who may not have any experience with electronic components. Now we get to the good stuff!

The place to get real radio kits you can use to actually talk to people is https://www.qrpkits.com/. Here you will find a ton of kits to do all kinds of things including keyers, voltage monitors, amplifiers, filters, and much more.

In case you are wondering, QRP technically is the Q code for "reduce power" but is widely used in amateur radio to mean very low power communications. It is considered quite the badge of honor to make contacts using extremely low power. You will hear more about that later in the book.

Repair vintage equipment

If you are an electronics person, or just someone who enjoys the challenge of bringing broken things back to life, then this might be the niche for you.

I just recently repaired an old HTX-404 handheld transceiver displaying an er1 code bringing it back to life with about an hour of my time and $10. It was nothing really to brag about, it is an easy fix, but the feeling of taking something old and broken and using your knowledge and skills to bring it back to life made it worth it.

To be honest, the HTX-404 I revived isn't that old, just from 1996. I have a Heathkit GR-64 Shortwave Receiver from around 1964 I did some very minor repairs to and it too is working again.

For those keeping score, that Heathkit was built well before I was born, and that is kinda cool.

The neat thing is that there is no shortage of broken radio, scanner, and other related electronics available for you to choose from. I like to do searches on eBay for shortwave radios, tube transceiver, or Heathkit.

Of course there are tons more things out there you can fix and use but those are great places to start. Right now that search for Heathkit resulted in 5,683 results. In the results you will find receivers, transceivers, power supplies, capacitor testers, and so much more.

This also has an added bonus in that you can repair items, use them, and then resell them to help pay for new items to fix and use. I met one ham who told me he spent an initial $20 on a broken radio off of eBay and ten years later he has repaired probably a hundred items and never had to pay for a single one as his repaired items paid for the new items. Nice.

I have to admit that although I love my new-fangled radios with all their new technologies and digital capabilities, that old equipment can be a lot of fun to play with as well.

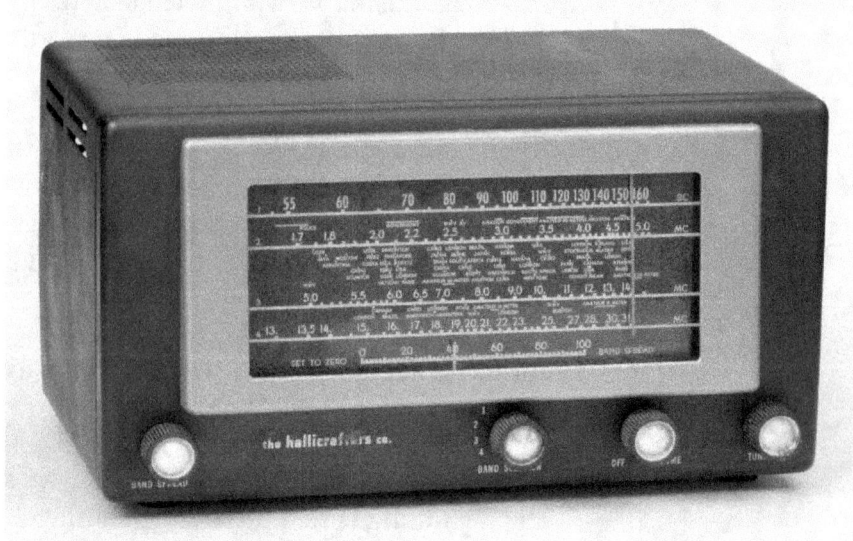

When I was a kid I somehow came into possession of an old Hallicrafters 5R10A shortwave radio and I still love to play with that radio over forty years later, although a different actual radio, it is the exact same model.

Build and test antennas

Everyone outside of professional radio guys and amateur radio operators talk about power. More power! She canna take it Captain, I' givin her evethin I got! Ok, showing my inner trekkie there. If you are a Star Wars fan, I'm sorry, not that I said that above, but that you are a Star Wars fan ☺

Anywhoooooooo......

We radio guys all know that power does matter, but not as much as the antenna. A better tuned antenna, a lower SWR, better matching, better location, higher altitude, it all makes a huge impact on the performance of your radio in both sending and receiving.

Most of us, however, do not live on a thousand acres with a 145' tower so we have to make do with what we have. This means we need to maximize everything we can to make what we do have the room and altitude for, do the absolute best job it can.

I have seen people put up antennas that look more like something out of Sanford & Son's reject pile than something I want hooked up to my radio only to find it worked spectacularly well.

From hoops to J-Poles, single wires to dipoles, verticals to yagis, the list goes on through a huge array of designs.

What's more, you completely design and build your own, or you can buy one ready-made and modify it. Whatever works best for you. The possibilities are endless.

You can even violate your HOA, city, state, and federal laws like the person probably did above!

Despite the pictures here, simple antennas can work well too! My primary HF antenna is a simple G5RV dipole suspended in the trees behind my house fed with ladder line.

This simple and hard to see antenna has reached all over the world including Saudi Arabia, Russia, all over Europe, South Africa, all over South America, and much more.

Contacts I have made have been on 10m, 20m, 40m, and a couple on 80m so the antenna works well on a wide variety of frequencies.

I think I paid about $80 for it and spent an afternoon getting it up and on the air.

For local contacts I am using a simple vertical, the Diamond X50A VHF/UHF dual band 2Meter/440 amateur radio base antenna. I mounted it to the back of my house and camouflaged it so it was hard to see using the technique outlined here: http://www.besthamradio.com/camouflaging-or-hiding-your-antenna-avoiding-antenna-restrictions/

See if you can find it in the image above!

To build a simple HF dipole antenna, see this guide: https://www.electronics-notes.com/articles/antennas-propagation/dipole-antenna/hf-ham-band-dipole-construction-80-40-20-15-10-meters.php or a simple Jpole antenna for 2m with this guide: http://www.k8qik.org/tech/pdf/jpole2m.pdf

If you decide to pursue antennas there are two books you should look at, both by the ARRL. Start with Basic Antennas.

And then move on to The ARRL Antenna Book.

Both are excellent books but I highly recommend the one on basic antennas unless you are well versed in antenna theory. Either way, grab some wire and have some fun!

Bake a pi!

The Raspberry Pi is a small computer on a board that has taken the world by storm. At first, thinking of actually learning to build and/or use another computer when you have your perfectly functional laptop or desktop seems silly. Read on to see why it might not be.

The Pi is small, just under 4 inches by 3 inches by just over an inch. In that size it packs four USB ports, an Ethernet port, HDMI, audio out, a micro USB for power, and both 2.4GHz and 5GHz wireless network. Different models have different specs, this particular model is the 3b+.

So, again, you have a computer, what good is this thing?

The first thing that pops into my mind is that this is a cheap computer that can do things like decode data modes, run SDR radios, track satellites, track flights, control antenna rotators, and even be a retro gaming machine.

Sure, your computer can do those things too, but it is nice to have a cheap replacement that takes up almost no room allowing you to use your main computer for other tasks.

So you think that is too boring? That's OK, we are coming to the good stuff.

My favorite use for the Raspberry Pi is to make it into a MMDVM hotspot. This allows you to get your DMR, Fusion, or D-STAR radio connected to the digital networks via the internet from your home, no repeater needed. How cool is that?

You can buy one ready to go like this SainSmart MMDVM, or you can build one yourself from scratch using this absolutely amazing Kindle book by probably the greatest author known to mankind (yeah, you know it's me, right?) called Building, Setting Up, and Using the MMDVM JumboSpot for DMR, Fusion, & D-STAR.

I won't be hurt if you just buy the one already built, that is what I did for my first one too. The rest (I own several, what can I say, I like pie!) I built myself.

You can also use them to make a WSPR beacon (more on that later) with the documents here:

https://github.com/JamesP6000/WsprryPi, and even make your own APRS Igate (more on that later too!) with the information here:
https://github.com/wb2osz/direwolf/blob/master/doc/Raspberry-Pi-SDR-IGate.pdf

Get an Arduino

We just got through talking about small computers but this one is a little different. Arduino is more input/output oriented and less computer oriented. It doesn't really have a computer operating system like you think of.

I think of a Raspberry Pi as a full blown computer in a small package. I think of this more as a programmable input/output appliance.

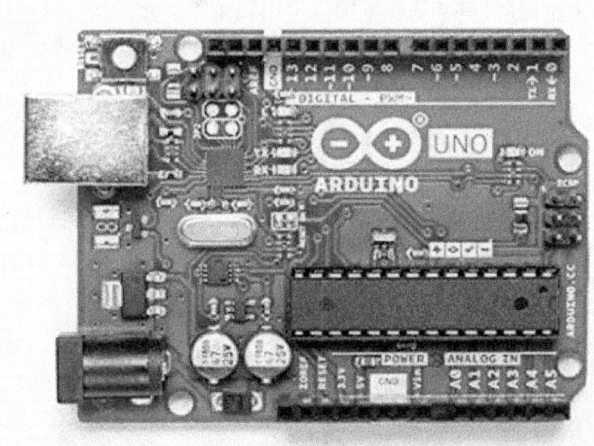

That doesn't mean the Arduino is not as capable, oh no, just different!

Some of the things you can build with one include an automated antenna switch, CW keyer, automatic antenna tuner, RF power and SWR meter, APRS tracker, beacon controller, and much more. Those projects can be found here:
https://www.dxzone.com/9-amazing-arduino-ham-radio-projects/

The idea behind this little miracle is that you have a power port, a programming port, and a whole bunch of pins that can be programmed to do a variety of things.

Examples are you can put various sensors on it and tell it when a sensor is activated or reaches a certain level, to output something on another pin.

For a really simple idea, you could put a small light sensor on a pen and program the Arduino so that when light no longer shines on that light sensor (it gets dark outside), turn on the voltage to another pin that has a light attached to it.

Sure, this is a really simple example, but considering you program the device using a simplified version of the C programming language and that it has a variety of inputs and outputs, including both analog and digital, and a few PWMs, the things you can make it do are just about endless.

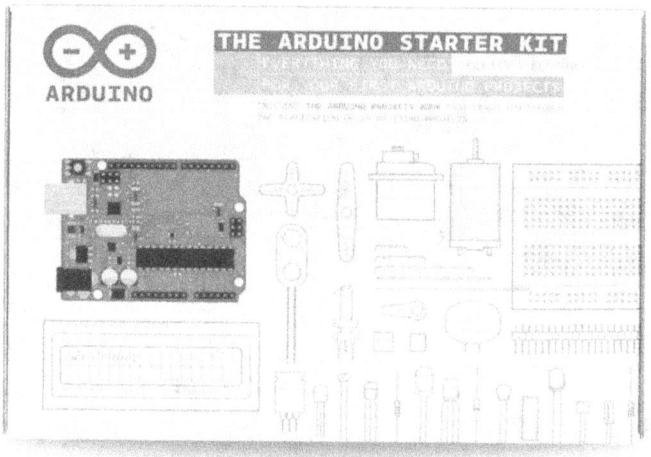

I personally like the Arduino Starter Kit as shown above as it includes easy to follow lessons and everything you need to complete the lessons (except for a computer to connect to and program the Arduino) right in the box.

If you haven't played with one, try it. If you really are an anti-social amateur radio operator, this should be right up your alley.

Help others with a blog or website

Just because you are antisocial doesn't mean you don't want to help others, and you can do that while sitting at home in your Psycho Bunny PJs working on amateur radio projects.

Moreover, you can use it to help yourself as keeping notes and images that you use to make your blog helps you remember what you are learning as you are doing it and gives you something to refer back to later. You can even do it for free.

If you want a free solution to start out with, head on over to https://wordpress.com/ and set up one of their free sites. See how it goes, if you like it and enjoy it, consider one of their paid options which removes the ads and adds more features.

If you decide to get serious, get your own domain name and host it yourself. Personally I use IONOS by 1&1 and have had good luck with them, you can visit them here: http://www.1and1.com/web-hosting?kwk=10598158&ac=OM.US.US930K244547B7030a

You can keep it an informal affair like James Stevens over at http://www.hamblog.co.uk/ where he posts all the neat things he is currently doing in amateur radio including transmitting from Snowdon, the highest mountain in Wales, in the snow.

Yeah, I'll take a pass on sitting in the snow on the top of a mountain, but it was awesome watching in his video.

You can also make it more of a formal affair with how-to articles, reviews, and recommendations like BestHamRadio.com:

I would wax poetic about what an awesome website that is and how amazing the guy who runs it must be, but then you would figure out it was my website and let out all my hot air. ☺

Those two websites will give you some ideas on things you can do, and how you want to do them. There is no correct approach, just make it how you want it and go from there.

There is little that feels as good as when you get an email thanking you for the article or suggestion, helping others is always awesome.

Things to do

Collect QSL cards

In today's digital world most of the contact tracking and verification is done by electronic means such as Logbook of the World (LotW). That's really nice, fast, and efficient, but it is also a little boring.

QSL cards are really neat because it is a physical card mailed from somewhere in the world you likely have never been and you never know what might be the image on the front of the card.

You can still get most people to send you QSL cards, but you need to make it brainlessly easy and completely cost free for them.

I typically send my QSL card as fast as I can in an envelope with another self-addressed envelope and a two dollar bill.

The Post Office sells international postage stamps you can use to get the envelope to your contact and then the two dollar bill is to them to pay for postage back.

I include a two dollar bill because it is unique and they will really appreciate the time and effort you put into the little package and are much more likely to send you a card back.

Be warned that probably 10-20% of the envelopes you send out may never be returned for various reasons but that is still up to 90% return on your investments so pretty good.

I have some really amazing QSL cards from all over the planet including special event islands, mountains, the Queen Mary, and much more.

For me, the contact is just the first part, receiving a physical object from the person, possibly on the other side of the world in a country I will never visit, is the second and most important part.

The image above is an self-addressed envelope I received back from a radio operator in Bulgaria complete with Bulgarian postage stamps.

Of course, your feelings may differ but that is what makes this hobby so great, so many different things for so many different people!

Chase awards

Chasing awards can go hand in hand with both collecting QSL cards and contesting, or not.

Different organizations offer awards for completing different tasks such as contacting people in every state in the United States (Worked All States), or for contacting people in 100 different countries (the DXCC).

There are a ton more of course, and the place you need to go to get started is http://www.arrl.org/awards. While you absolutely can get awards from other groups, this is normally where everyone starts as they have the most recognized awards around.

Today the way most of this works is like this:

You make a contact and enter it into your Logbook of the World account. Your contact does the same and that verifies the contact.

Once you have the required amount and type of verified contacts (for example, a verified contact in each of the 50 US States) you apply for the award. Your contacts are verified through the award program and then your award is issued. Simple!

Except that getting some of these awards is anything but simple and in fact can be downright exhausting. It never fails that when you get down to one or two left, it seems that everyone in that state, country, or continent has given up the hobby and you never hear them.

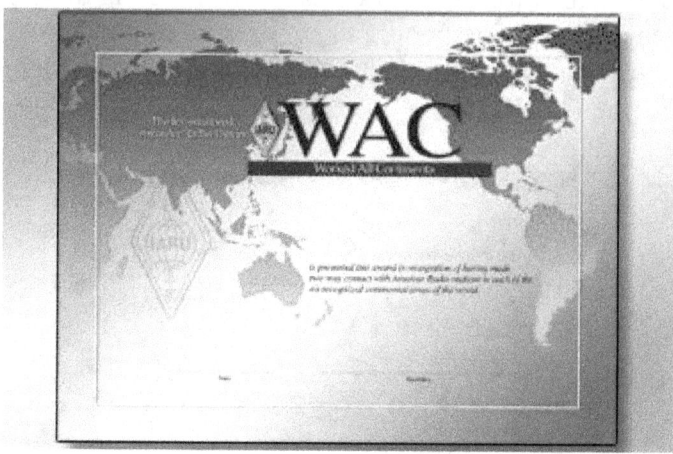

The funny thing is that while you may wait months for that contact in Maryland, other people have a hundred contacts in Maryland and none in Ohio for which you have thirty contacts. It's just insane sometimes which makes the hunt that much more fun, at least to me.

Radio astronomy

Radio astronomy isn't transmitting so technically you do not need an amateur radio license to participate but having the radio background sure makes it easier.

So what exactly is radio astronomy? Think of it like using a telescope to look at the moon, except you are using an antenna and radio to listen to the moon. So the moon really doesn't have much to listen to, but you get the idea.

Most people's first target is Jupiter which absolutely has a lot to listen to. In fact, you can even listen to the interaction between Jupiter and its moon Io. Amazing.

Probably the best place to start is with https://www.radio-astronomy.org/, the Society of Amateur Radio Astronomers. There is a wealth of information and links here that is sure to get you going in the right place.

If you are like me and prefer to just jump into the deep end without all the reading (you know the type, those who never read the instructions until after you can't get it to work) then you might just head straight to building a radio telescope with the excellent article http://www.arrl.org/files/file/ETP/Radio%20Telescope.pdf by Mark Spencer that was published in the June 2009 QST magazine.

The most popular home brew radio telescope project I have found is the Itty Bitty Radio Telescope which can be found at http://www.setileague.org/articles/lbt.pdf, and David Haworth's excellent site here: http://www.stargazing.net/david/radio/itty_bitty_radio_telescope.html.

That's right, use that old satellite dish you still have on your house from before you went to streaming for something useful, turn it into a radio astronomy dish and "listen" to the sun, Jupiter and more!

FT8 digital

If you have an HF rig and want a different way to communicate with people all over the world, try a digital mode like FT8.

Digital modes can be a little confusing to the newcomer because when people use the phrase "digital mode" they could be talking about something like FT8 which is much like the old modem over a phone line digital communications.

They could also be talking about digital modes such as DMR, Fusion, and D-STAR which are ways to communicate using the internet and/or repeaters.

With protocols such as FT8 you still make connections the "old fashion" way with an analog radio and antenna, relying on propagation and weather. The difference is that you type your messages out using a computer.

Depending on the software you use, messages may be semi-automatic as well. This leads to some people calling this "cheating" when it comes to communications but those are probably people sitting with their old tube transceivers using nothing but CW, which is fine, but their beliefs do not at all impact the rest of the world that may disagree.

The most popular software right now is WSJT-X which is available from Princeton here: https://physics.princeton.edu/pulsar/K1JT/wsjtx.html. The following image shows the interface of the software, it is more or less just point and click to send and receive messages.

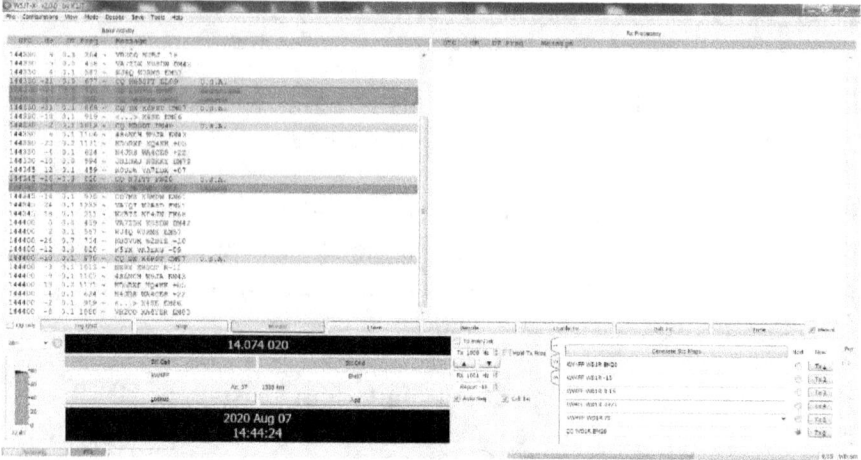

The way this works is that your radio is connected to your computer which can then send and receive information through the radio. As soon as I explain how this is accomplished it will be outdated so I will not even try to keep up with technology but will give you a couple of popular examples.

With a slightly older radio such as my Kenwood TS-570D, there is a cable that connects the radio to a serial port on my computer (or a serial to USB adapter). I then run a piece of software on my computer that controls the radio.

In some cases, depending on the software and protocol, you may need to use a different method which uses a SignaLink audio processor which connects between your radio and your computer. This acts like the modem in the old modem and phone line setup I talked about earlier.

Newer radios such as the Kenwood TS-590SG connect directly to the computer via USB and have the capabilities of a SignaLink built right in.

Most of these contacts are very short, much like contesting, you exchange minor information such as signal reports and maybe locations, then off to another contact you go. No muss, no fuss.

28

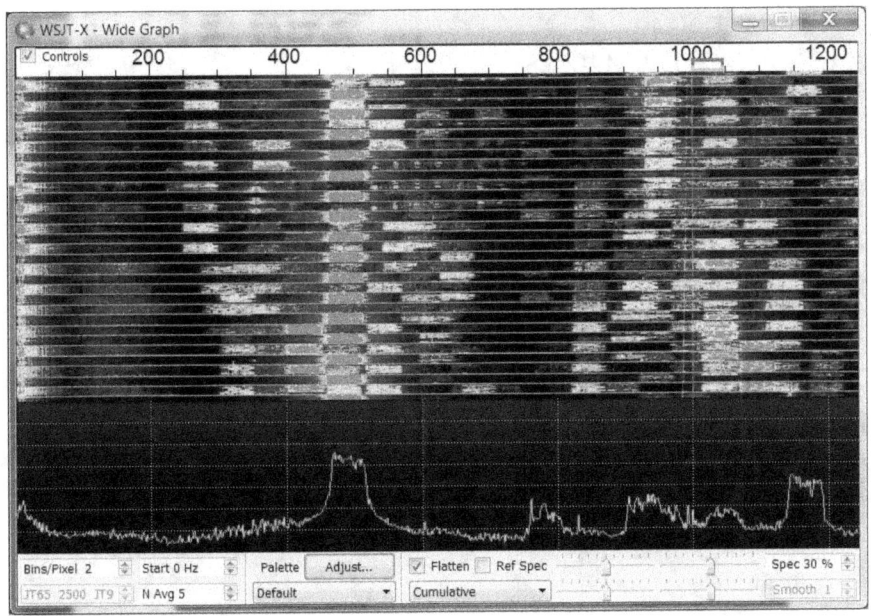

Many people can share the same frequency as shown in the two images on this page. This is done by using different sets of audio frequencies on the same radio frequency. It can also be done because one set of people send while another set receives.

The basic idea is that one station transmits their CQ, this takes fifteen seconds. Then a station replies to their CQ, this also takes fifteen seconds. This cycle is repeated until the conversation is complete at which point the original station calls CQ again and the whole process starts over.

The only information that is typically exchanged is call signs, grid locators, and signal reports. Conversations might appear like this one did:

CQ KW4FF EM97
KW4FF WD1R EM20
WD1R KW4FF -15
KW4FF WD1R R-15
WD1R KW4FF 73
KW4FF WD1R RR73

CQ KW4FF EM97
And so on....

On the first line, KW4FF is calling CQ and provides his grid square of EM97. I reply in line two with my grid square EM20. Then he replies with a signal report to which I reply R-15 or roger and my signal report. Then he sends 73 to complete his side of the exchange and I reply RR or roger roger, 73 to complete the exchange.

Then the whole process starts again. A complete exchange takes about a minute and a half to complete making this a very fast process.

You can see clearly in the display in the first image in this section that the country of origin is displayed on the right side of the message, making this the ideal mode to contact other countries you have never contacted. You can of course do the same with states although you have to manually decode the grid squares yourself but that gets pretty easy once you start.

SDR dongles

SDR is another activity that doesn't require a license to participate although having the license and listening to SDR complement each other.

This is another place where terms can get mixed up. SDR technically stands for Software Defined Radio and refers to a radio where most or all of its functionality is defined by, you guessed it, software instead of hardware.

There are any number of SDR radios for a variety of functions ranging from cheap little receivers (what we are talking about here) all the way up to very nice HF rigs.

What is so special about the little dongles we are discussing here is not only that they are cheap but they are extremely flexible. Many of them can be used right out of the box from 25MHz up to 1700MHz. Use of add-on kits can extend that down to around 100kHz. You can hear a lot of interesting things with that!

You can use it as a scanner for any frequency you want, even those that are not in traditional scanners.

The thing I really love is that they are cheap! You can get a really cheap one like the Nooelec NESDR Mini 2+ for about $25, step that up to a much nicer one like the NooElec NESDR Smart v4 for under $40.

You can extend your capabilities with the power to transmit on virtually any frequency using the HackRF One which allows you to do just about anything with radio you can think of including recording your garage door opener's output and playing it back at will.

The HackRF One even has a set of excellent videos that use the device to teach you a ton about software defined radio and what all you can do with it.

Once you get the dongle you need the software. In this example I am using the freely available CubicSDR available from https://cubicsdr.com/

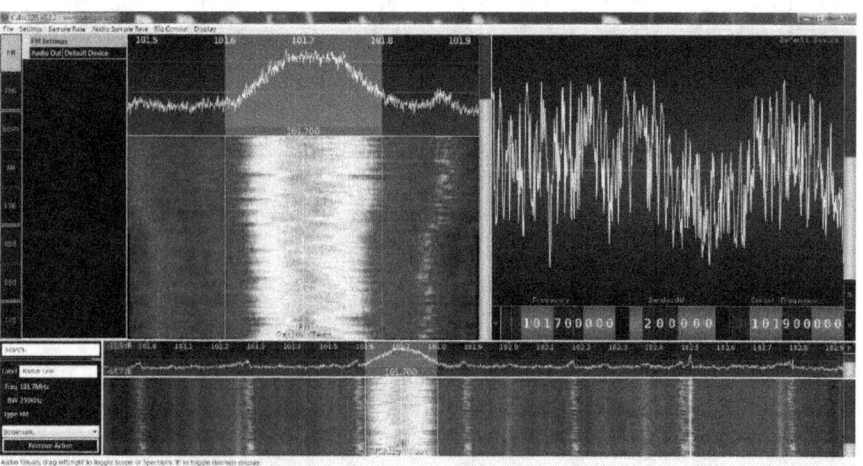

In the above image I am listening to a local FM radio station.

Pirate hunting, arrrr

There are those individuals out there who want to operate without a license and generally cause mayhem. You can track them down with your very own pirate hunting adventure!

In amateur radio, we often have "fox hunts" where we place a low power beacon somewhere hidden and then "unleash the hounds" to go find it. You can use your skills to basically do the same thing by finding "pirate" stations operating without a license or in an illegal manner and "flush them" (report them to the authorities).

You might hear an obviously bogus radio station broadcasting pop music, or someone transmitting on the ham bands without using a call sign, or even someone on a CB cursing and yelling. Whatever it is, track it down and report it!

Before we go any further, no, you should never ever attempt to confront someone or even let them know you are looking for them as they have already shown they don't follow the laws and you have no idea how far outside the law they will go. Better safe than sorry.

To start with you need a receiver, this can be a handheld radio like a Baofeng UV-5R, and an antenna such as an Arrow Fox Hunting Loop as shown in the next image.

The Baofeng radio can receive not only 2m and 70cm ham radio bands, but also from 65MHz-108MHz in the standard FM radio band.

For more coverage you could use something like a [Bearcat SR30C handheld scanner](#) which can cover from 25-512MHz. It can also receive both AM and FM signals.

Finally you could use a computer and SDR like those we previously discussed for the ultimate in receiving capabilities. This method is also useful as you can record the signal being broadcasted.

QRP

For a serious challenge you can try QRP which is basically operating on the least amount of power possible. With most people I talk to anything less than about 5 watts is QRP with serious devotees working with around 1 watt of power.

You are probably thinking this sounds boring, I can grab a $25 handheld radio and be working QRP, particularly if I use it on the low power setting.

Yes, and no.

That is technically correct, but the idea here is to make contacts far away, as in another state or even country.

The typical HF radio that you use to talk to other states and countries puts out 100 watts, try that with just one watt of power! Then make it even more challenging and try it mobile.

One facet of QRP work is hiking trails, particularly up hills and mountains, and working people far away with just what you carried with you. This includes battery packs, portable antennas, the radio itself, etc. This gets interesting in a big hurry.

More information can be found at the QRP Amateur Radio Club International https://www.qrparci.org/ including lots of informational links.

Looking here you will see they even have awards much like the ARRL ones we looked at earlier in this book except these are all done with extreme low power.

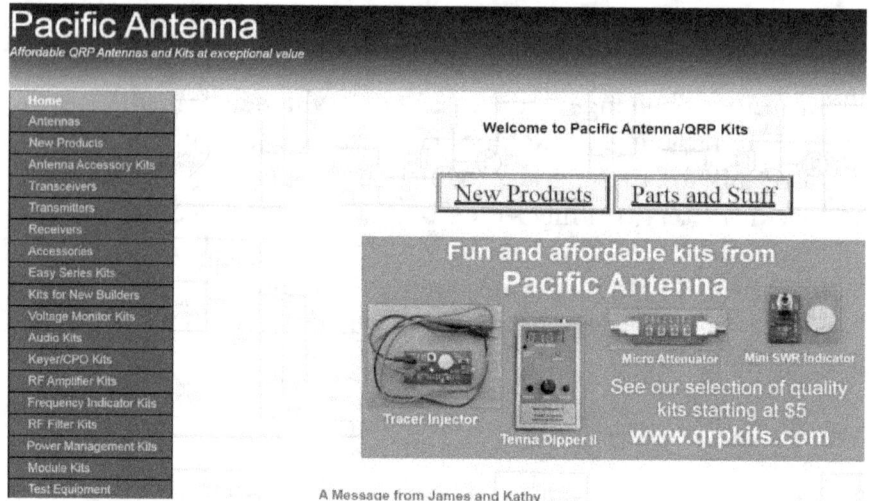

Of course as I mentioned earlier in this book, Pacific Antenna https://www.qrpkits.com/ has some nice QRP radio kits you can build yourself along with a ton of other stuff which is just amazing.

Working satellites

Yup, you got that right, you can use a satellite as a flying repeater in space. Using that repeater you can talk to someone next door, or someone in the next country over, just as easily.

The last time I looked there were dozens of satellites you can use and of course you can also communicate with the International Space Station at certain times too.

The typical setup is a dual band radio (2m and 70cm) connected to a specialized antenna like the Arrow II Dual-Band Handheld Yagi which you manually track across the sky. Other antennas can be used as well, the more directional the better, although on extremely rare occasions people have made connections with simple antennas such as a magnetic car mounted stick.

There is a little more to it than that, for example adjusting for Doppler shift which causes the frequency to shift as the satellite moves towards you, and again when it moves away from you. This is typically done by setting a range of frequencies in the memory of your handheld and switching between them as the distance and direction of the satellite changes.

Many of the satellites have a frequency on one band for uplink, such as on the 2m band, and another frequency on a different band, such as 70cm for the downlink. This is why most people use a dual-band handheld for this use.

Tons of information can be found at the Amateur Radio in Space website https://www.amsat.org/ as well as a great article by Sean Kutzko KX9X at https://www.onallbands.com/satellite-basics-part-1-guide-to-ham-radio-satellite-operating/.

If on the other hand you want to have the distinction of making contact with someone who isn't even on the same planet as you, check out https://www.ariss.org/contact-the-iss.html for great information on contacting the astronauts on the International Space Station.

Not only can you try and communicate with the astronauts but you can communicate directly with the ISS itself directly using packet or use its packet repeater. They also transmit SSTV on 145.80 on occasion.

Moon bounce

Also called Earth-Moon-Earth, is exactly what it sounds like, bouncing a signal off of the moon and back to a point on Earth. This has the distinction of being the longest communications path that can be used to communicate with someone else here on Earth since the moon is almost 239k miles from us.

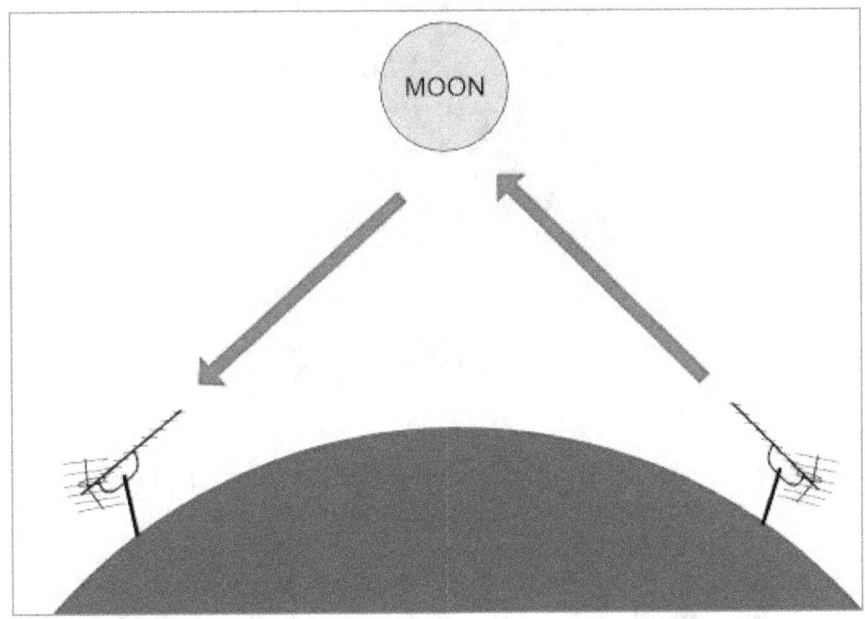

Moon bounce communications is typically done on 2m, 70cm, or 23cm. Many contacts are done using CW, digital modes like JT65, and occasionally using voice.

The majority of affordable moon bounce contacts today are done with a 2m mobile or base radio that can output 50w or more, a highly directional yagi antenna delivering a minimum of 10dBi gain (preferably more), and a computer running software for the JT65b digital protocol.

A good intro into moon bounce can be found at:
https://www.dxmaps.com/jt65bintro.html

If you are really interested, and how could you not be, there is a great Kindle book by Stephen Appleyard and Philip Malme called Getting started in EME: Earth-Moon-Earth Transmission.

This is also one aspect of the hobby where you can just go absolutely insane with massive antenna arrays, kilowatt amplifiers, and advanced trackers with antenna rotators to keep the moon right in your crosshairs. Fortunately you don't have to start out like that!

Meteor scatter

Meteor scatter is basically the same as moon bounce we just talked about except you are bouncing the signal off of the ionization of the atmosphere caused by meteors entering the upper atmosphere instead of bouncing off the moon.

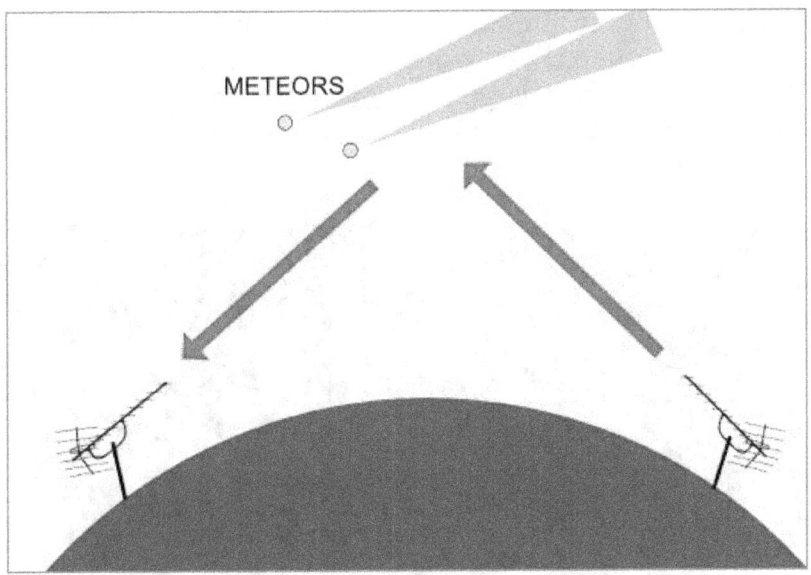

Meteor scatter communications is typically done on 2m with a few on 70cm. Most contacts are done using JT65b with an occasional CW contact. I have heard of people doing voice but never actually seen it successful as the time you have is too short.

Just like moon bounce, the majority of meteor scatter contacts today are done with a 2m mobile or base radio that can output 50w or more, a highly directional yagi antenna delivering a minimum of 10dBi gain (preferably more), and a computer running software for the JT65b digital protocol.

Unlike moon bounce you have mere seconds to make your contact as the ionization caused by most meteors is extremely short lived.

While meteors are constantly entering our atmosphere and ionizing it, those small meteors might only cause ionization suitable for a fraction of a second to a few seconds. Where the real fun lies is during a large meteor shower as these larger objects cause ionization that can last more than long enough for a JT65b contact.

The other advantages of working during a meteor shower are of course the light show, the fact that many other people will be working the mode thus significantly increasing the chances of contacting someone, the most significant the ionization, the more of the signal is reflected so your equipment instantly gets increased capabilities.

Slow scan TV

You can send and receive video (kinda) over the airwaves next door, or around the planet with proper propagation. In fact, SSTV has been used to send video TO the planet Earth by the Russians in 1959 on Luna 3 in orbit of the moon. It was also used by NASA to send images from the moon.

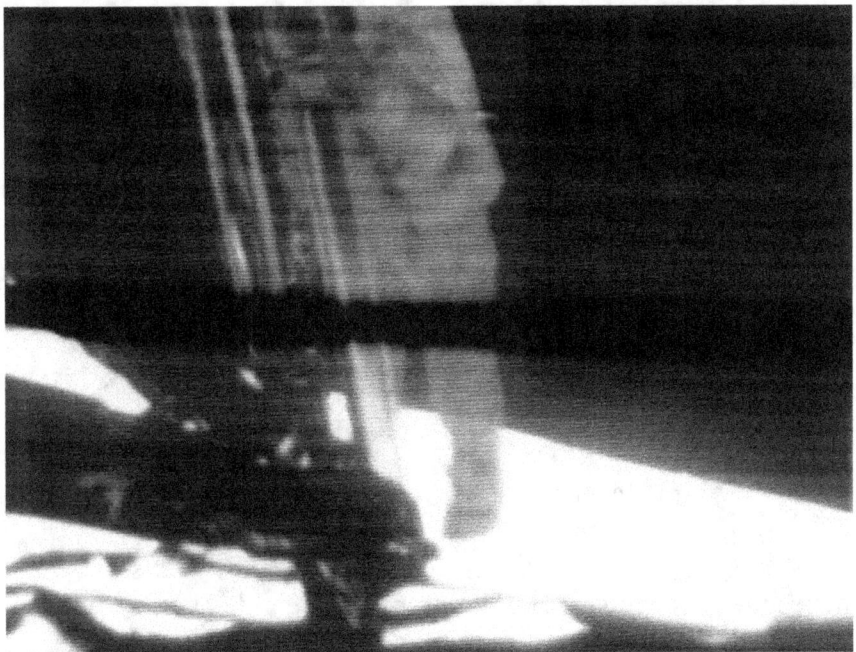

The basic idea is the same as many other digital modes such as the FT8 we already discussed. Your computer sends data through a sound card and into your radio which then transmits those tones to a receiving station or stations that reverse the procedure to view the data, in this case video.

As of August 2020 the ISS still sends SSTV images down as well. Check https://www.ariss.org/ for their schedule.

One of the best introductions I have found into receiving SSTV is by Ham Radio Concepts at https://www.youtube.com/watch?v=ojTcM2NGO48 and is certainly worth a watch.

Now one thing to remember is that this "video" is very slow, hence the term slow scan. Each frame of the video takes a little over a minute to appear completely on the screen.

No, you are not going to video conference with this technology, but you can set up a "web cam" to show activity going on if you like.

Or, if you are more of a voyeur then you can just watch what others put up all day long. ☺

Amateur TV

Yup, another TV mode! This time however we are talking about what you probably think of as television, actual video and audio that looks like what you see on your actual TV (or when you stream on your devices).

OK, maybe not that good, but you get the idea. Whereas we previously talked about SSTV which is one frame of the video every minute or so. Amateur TV is also called Fast Scan TV or FSTV.

A great little write up on ATV was written by Bill Munsil, K1ATV and can be found here:
http://www.hamuniverse.com/atvfastscantv.html.

You can also head over to http://www.hamtv.com/ and see examples of all the equipment you need to make your own ATV station.

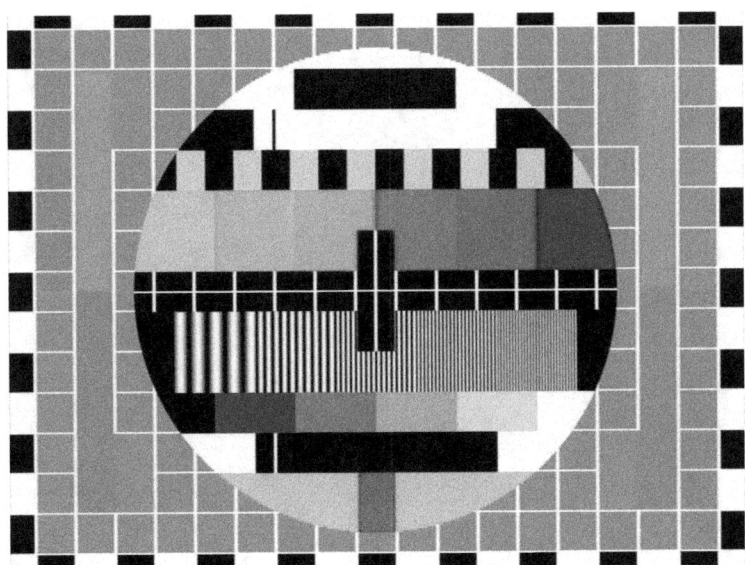

Recently there has been a large push into digital ATV making the difference between commercial TV and what you can do with an amateur radio license even smaller. A great intro into the digital side was written by Jim Andrews and can be found here: https://kh6htv.files.wordpress.com/2018/08/an-45-dtv-book.pdf

Of course you can't transmit ATV "for profit" or in any way that could be considered commercial, but you can certainly do it for fun, education, or experimentation.

Receiving on the other hand is cheap, easy, and you can do it with virtually no equipment you don't already have except a downconverter to change the received signal from what you receive on your radio antenna to something a typical TV cable input can use.

Kite antennas

Looking for something really unique? Try attaching your antenna to a kite. Make a 10m or even 20m full wave vertical!

Like most things, it isn't quite as easy as it sounds. You don't just grab any old kite, attach the antenna wire where the kite line should be and start flying, no no no.

First off the kite should be one designed for lifting and it should be a single line kite. I have seen this: https://amzn.to/3a7BFBT recommended on a couple of sites.

Next you need to attach your kite line to some flexible line or cord several feet back from the kite, probably 10-20 feet of the flexible line should be enough. Then that line is attached to an isolator and finally to the antenna wire which is then secured to the ground directly under the isolator so that it makes as much of a vertical antenna as possible.

Then you fly the kite and attach it to something so you do not have to hold it while you work the radio.

This is an excellent idea for QRP as it maximizes your antenna to really get your signal out there. In addition, most of the equipment you need packs up very tightly so it is less to carry into the field.

One of the best beginner resources on antenna kites is SOTABeams at https://www.sotabeams.co.uk/kites-faq/. I also like the article at https://www.sbarc.co.uk/club-talks/talks-2013/kite-antennas/.

The history of using this technique dates back to at least 1905 in the United States Signal Corps and went on to be widely used in World War II first by the Germans and later copied and used by both British and US forces.

Weather FAX

Commonly called WEFAX, Weatherfax, or HF fax, this can allow you to see real time weather information and satellite images from just about anywhere in the world where you have a HF radio and a computer.

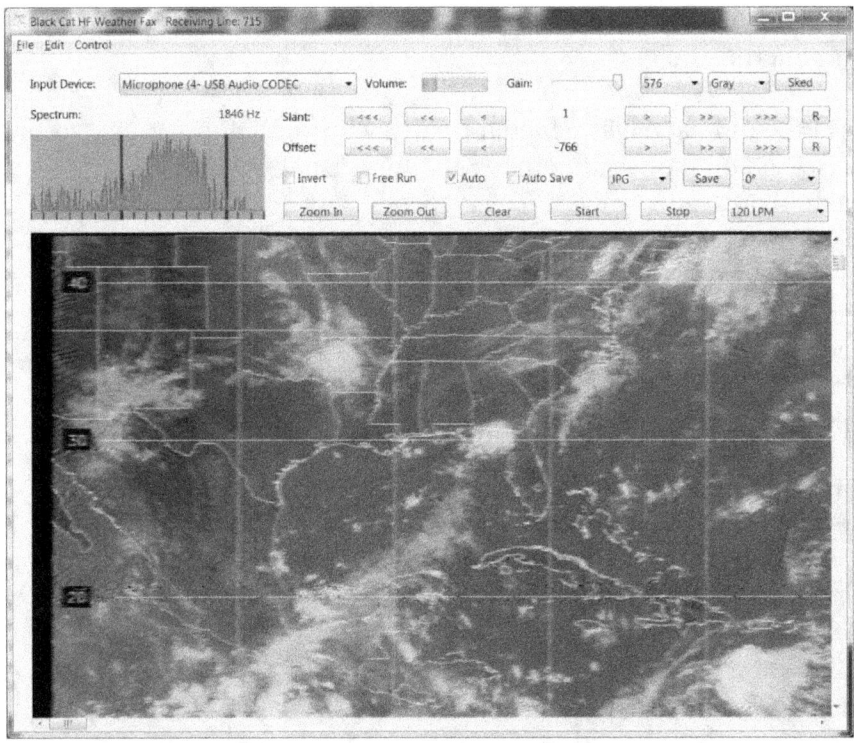

Above is an image of me decoding a satellite image of current weather over my part of the world using my HF radio, SignaLink audio device, and a piece of software called Black Cat HF Weather Fax available here:
https://www.blackcatsystems.com/software/hf_weather_fax.html

There are quite a few pieces of software that can do this but this one is almost too easy to set up and use, can automatically save the faxes as they come in so you can view them later, has a fully functional free trial, and only costs $20 when you are ready.

So why in the world would you want to receive these slow, black and white weather documents? I can think of several reasons, lets start with one I have personally used…..

During Hurricane Ike my power was out for about two weeks. Not just mine, but all the power in the whole area, which included my internet service provider and all the cellular towers. No internet at all unless you had satellite internet, which is basically no internet, heh.

What I did have was generator/battery power and my radio, so I also had all the weather maps and forecasts I wanted, automatically.

When at sea you may not have internet, and may not want to spend a small fortune on satellite phones. With these faxes coming in automatically you can check the weather any time you want to by opening the directory on your computer and viewing the latest files.

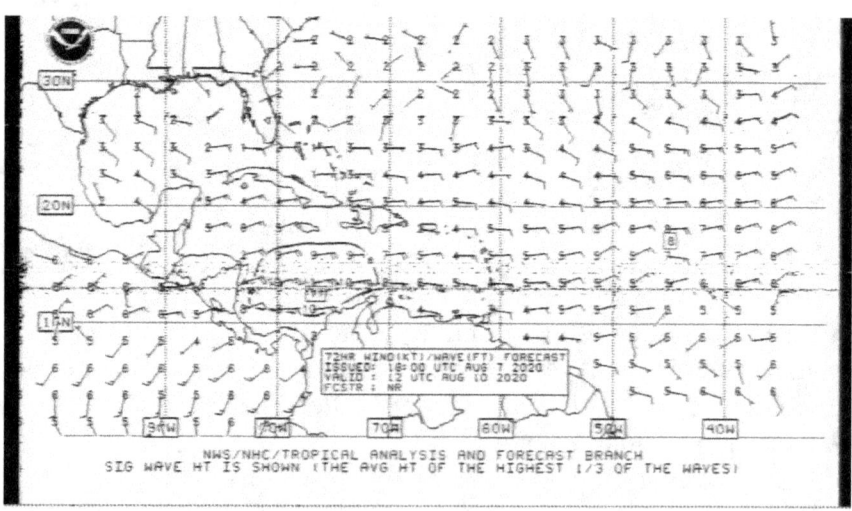

Here is an actual weather map that the software saved automatically.

QRSS

Morse code is one of those things where people either want to do it, or have no interest in it. Some of those who don't want to learn it would still like to participate and that has led to computer programs that encode and decode Morse code, or CW as it is more commonly called in amateur radio.

I have yet to see a setup that was really good at it, maybe good enough that you got the idea of what the person was trying to say, but not good enough to really carry on a conversation under real world conditions. Maybe I just haven't seen it yet.

One of the reasons people are so interested in CW is that it can get through when nothing else can, which makes it extremely desirable for long distance communications.

A solution to that is QRSS which is extremely slow CW. This is typically accomplished by home-brew hardware solutions although you can use standard hardware and software.

To view QRSS the most popular software out there is ARGO available for free at: https://www.i2phd.org/argo/index.html

This software allows you to "see" the Morse code and decode it.

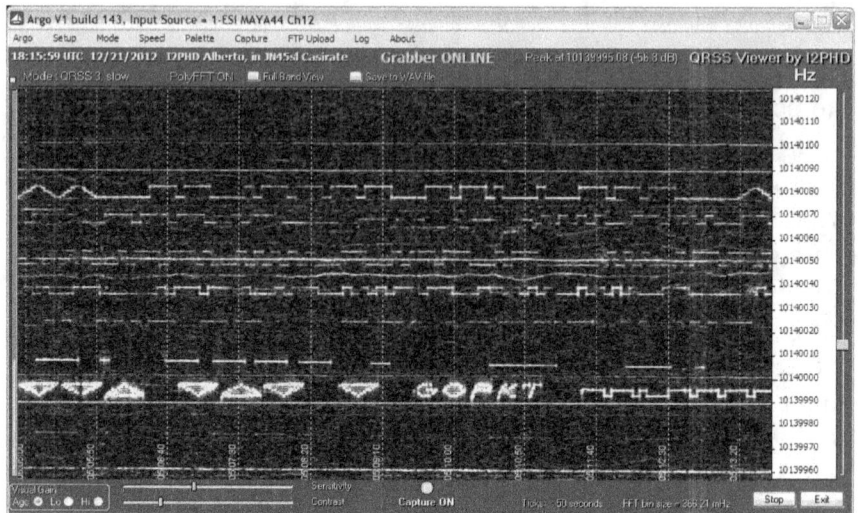

Another common program to use with QRSS is Spectran which is available here: https://www.sdradio.eu/weaksignals/spectran.html

More information on QRSS can be found on QSL.net at https://www.qsl.net/m0ayf/What-is-QRSS.html.

SOTA summit hiking

So you want to do something with amateur radio that is really antisocial? Take your radio to the top of a mountain!

SOTA stands for Summits on the Air and is an organization that promotes portable operation in mountainous areas according to their website which can be found here: https://www.sota.org.uk/.

I have to admit I know little about this aspect and am not a member of SOTA as I am not about to drag my rear end up a mountain with a backpack full of equipment, but that is just me. I have nothing but awe for those who do.

They also say they have activities for people who stay at home and are operational in over a hundred countries around the world. I might have to give that a shot.

The idea here seems to be an "activator" climbs a hill or mountain carrying everything they need to run their radio "station" on foot or on a non-motorized mountain bicycle. You then make four contacts with "chasers" and earn points. You can also only use a hill or mountain once per year.

These points accumulate to get you awards, even chasers can get awards for contacting enough activators. In fact, from my reading you can participate in all this with just a receiver by logging the contacts between activators and chasers so everyone can join in the fun.

There is a nice beginner's article by Bob Witte up on the ARRL website at: http://www.arrl.org/radio-operating-from-summits and a really nice podcast covering it on the W0C SOTA website at: http://www.w0c-sota.org/about-sota/.

It seems that people are either using a nice dual band radio like the Yaesu FT-60R for just VHF/UHF or a Yaesu FT-818ND for HF. Antennas vary considerably and seem as much about preference as anything else including nothing more than a simple extendable antenna to dipoles hung from the trees.

Both of the radios listed above have battery packs included so power isn't a problem although I have seen mention of solar panels like the BigBlue 28W that people strap to their packs to charge the batteries on multi-day hikes.

APRS

APRS stands for Automatic Packet Reporting System and is a way to automatically send data from your radio. If this sounds obscure, well it is. But that also means it has some really neat uses.

In particular one very useful aspect is when used with a portable radio that has both APRS and GPS is that you can automatically report your location at regular intervals.

What good does that do? Well take a look at this:

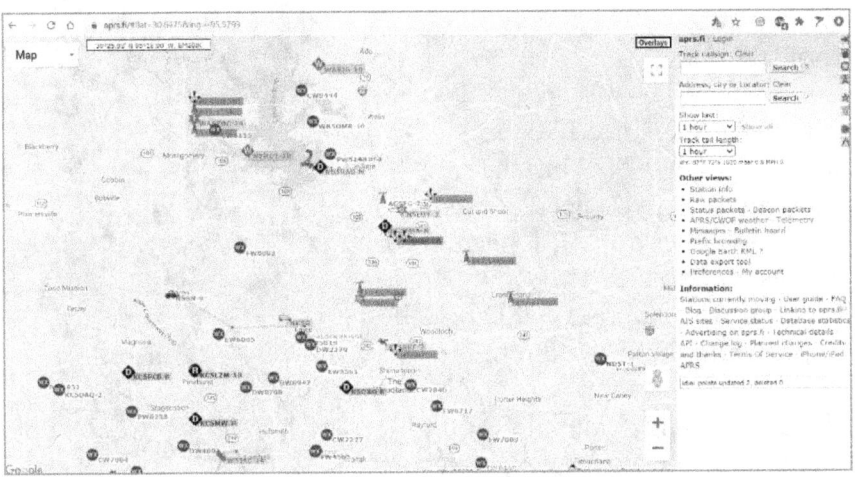

That, my friends, is a Google map with APRS data overlaid that is available at https://aprs.fi/.

So how does information get from your radio to that map? Glad you asked!

First, your radio needs two things, a TNC which is a modem for radios. This allows it to send and receive data instead of, or in addition to, voice.

Next we need a GPS so that the radio knows where it is in order to send that data.

All of this can be put together a couple of ways; by having everything built into the radio, or by connecting the radio to a TNC and/or GPS somehow.

Radios that have all this built in include the Kenwood TH-D74 and TH-D72, the Yaesu FT-1DR, FT-2DR, FT-3DR, and VX-8GR (the VX-8DR can do it with a plug in GPS module). These are the only analog APRS radios I know of.

Digital APRS radios like the Alinco DJ-MD5TGP but the problem is that you need constant access to either a DMR repeater or hotspot like the MMDVM which is far less practical than a single analog radio solution.

You can also do some interesting things like take a cheap Baofeng UV-5R, attach an APRS cable and plug it into an Android smartphone using the headphone jack (assuming you still have one of those).

However you get it connected, the idea is that the setup sends out a packet of information on 144.39MHz which is received by either digipeaters or Igates. If it is heard by a digipeater it is repeated back out a certain number of times before pretty much being dropped.

Hopefully the packet will be picked up by an Igate which is an internet gateway. Once this happens, the information is sent to the internet to be displayed on maps like the one you saw earlier.

If you look on the map you will see things like black stars and green stars, these are digipeaters and Igates.

You will also see a lot of blue circles with WX inside of them, these are weather stations. You can click on them to get more information possibly including wind speed and direction, temperature, humidity, barometric pressure, etc. Not only that, but on some you can click on the link for a weather graph and see the history of the weather at that location!

Some of these weather stations are just off-the-shelf setups like the Davis Instruments 6152 Vantage Pro2 as shown to the left which is supposed to be able to connect to a TNC via their data logger cable and send out APRS data. The instructions from Davis Instruments are available here: https://www.davisinstruments.com/product_documents/weather/app_notes/AN_35-ham-radio-for-aprs.pdf

Several people report that you can use a radio like the Kenwood TM-D710G to directly connect such a weather station since it has a built in TNC.

You can even make some out-of-this-world contacts by using the digipeater on the International Space Station! There is a great article by JoAnne Maenpaa located here:
http://tinyurl.com/K9JKM-ISS-Packet-Radio

Lots more information can be found in a great article and video my Michael Martens located here: https://www.jpole-antenna.com/2018/09/17/introduction-to-aprs-the-automated-packet-reporting-system/, in this article on the wb8nut website http://wb8nut.com/aprs/ and on the ARRL website at: http://www.arrl.org/aprs-mode#:~:text=Long%2Dtime%20packeteer%20Bob%20Bruninga,%2D%20and%20text%2Dtransfer%20activity.

Sorry about that huge link above, seems the ARRL needs to figure out shorter links.

Other aspects of APRS include the ability to send something akin to a text message across the network to another user, sending radio direction finding bearings, telemetry data, and more.

Like much of the other things in amateur radio, your imagination is the limit!

WSPR low power beaconing

Weak Signal Propagation Reporter, pronounced whisper, is basically a set of low power beacons that anyone can set up or just listen to. The idea is to use these beacons to check propagation conditions on various bands around the world.

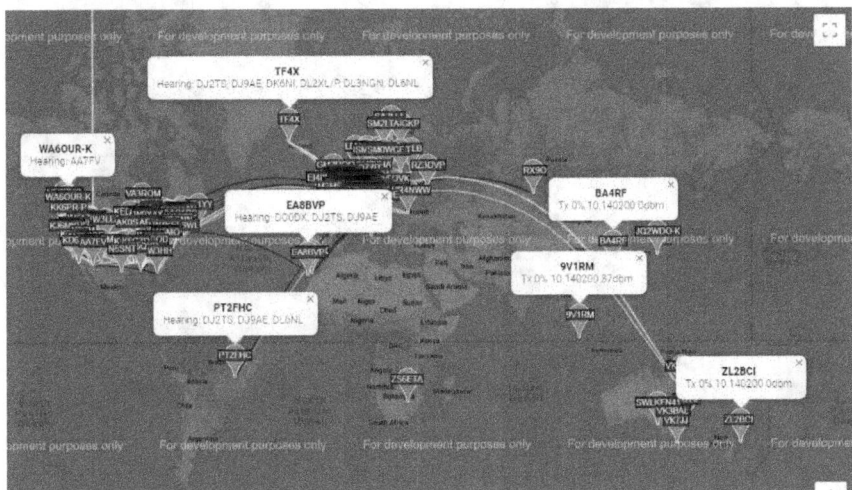

The map above shows some of the data from http://wsprnet.org/. Note on the pop-up white boxes it shows the station ID and either who they are hearing or statistics about their transmission.

You can either transmit or receive the data directly by using the WSPR software available from:
http://physics.princeton.edu/pulsar/K1JT/wspr.html

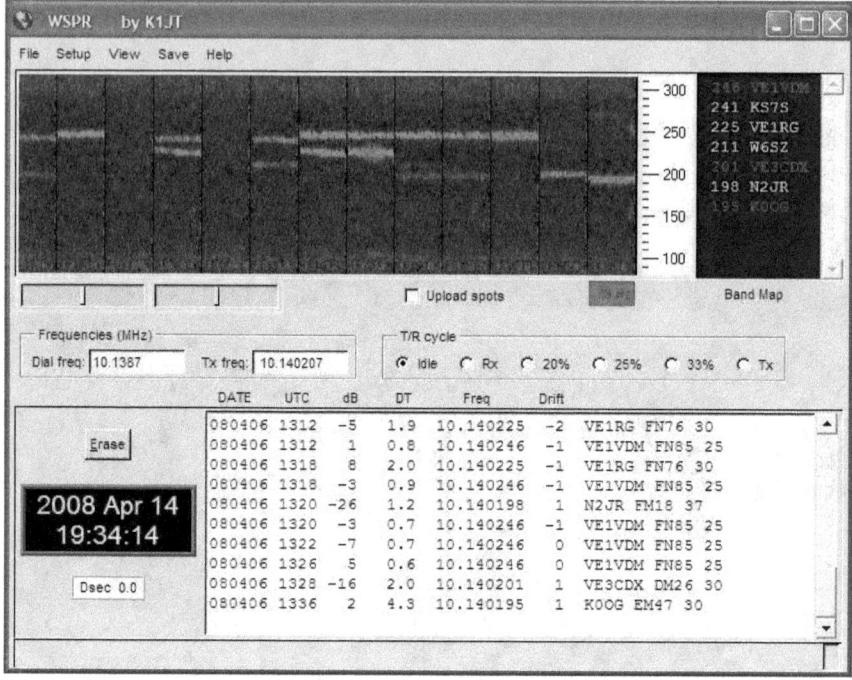

Even if you never transmit, this will allow you to see what the band conditions are on a given frequency. Use this information to plan your HF activities, study the weather patterns, whatever you want.

Heck, you can even connect your Android phone as a beacon assuming you have a headphone jack that you can connect to the audio input source of your radio, then set the radio to VOX and away you go!

Things to participate in

Contesting

Just like amateur radio as a whole, contesting has several different aspects you can participate in. You can run a contest station, support the station, contact contest stations, and much more.

I have enjoyed unofficially participating in many contests over the years and had a tremendous amount of fun doing it. Basically I spend my time trying to get stations to return my call, trade a little information, and then sign off from that station, all very rapidly.

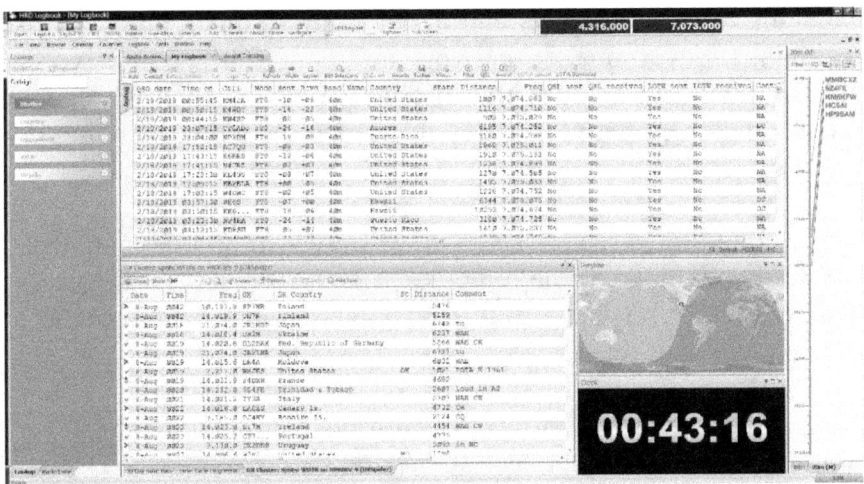

Shown above is a typical contest log showing contacts in the top portion of the screen, available stations that have been spotted on the bottom area, and of course a map and clock on the bottom right.

A typical contact might sound like this: I hear a ton of people calling out their call signs trying to get the attention of someone with the call sign of KC5VQH:

Me: WD1R.........WD1R............WD1R
Them: Ending in 1R
Me: Whiskey Delta 1 Romeo

Them: Whiskey Delta 1 Romeo, KILO CHARLIE FIVE VICTOR QUEBEC HOTEL, 55
Me: 55, 73s
Them: 73 QRZ

All of this took mere seconds, and the cycle repeats for what seems forever. The rarer the person (in a country with few radio operators, a special event station on an island, etc), the harder it will usually be to get them to recognize you over the other kabillion people calling. The guy in Detroit? Easy.

Let's break this down a little. The first line is me calling out my callsign. You repeat it over and over but not so much that you are stomping on everyone, or that you can't hear if they returns your call. You will figure it out after listening a while, usually I repeat mine somewhere between every five to ten seconds depending on the situation.

The second line is typical since they rarely hear all of someone's call so they will say something like "starts with" or "ending in". This very often results in multiple people replying to him and then he will pick one.

Next is them replying with your call followed by their call. Sometimes when the connection is good you will hear "WD1R, KC5VQH, 55". Sometimes when the signal is poor they will use the full call phonetically with "Whiskey Delta 1 Romeo, KILO CHARLIE FIVE VICTOR QUEBEC HOTEL, 55".

The 55 tagged on the end is a signal report and honestly I think I might have heard something less than 55 a couple of times but it is almost always 55 just for expediency. If they hear and understand you, they send 55, period.

I then reply with my signal report of 55 and then tell them goodbye with 73s.

Lastly they reply with 73 and QRZ which stands for "who is calling me?" and the whole process starts again with a ton of people giving their calls trying to get the station's attention.

In this scenario there could be hundreds of station operator type people and thousands of people like me trying to communicate with them. The station operators are usually contest stations or serious contesters who are entered into the contest and want to win.

Do I have to enter the contest to talk to him? Absolutely not! You can just jump right in and have fun!

The most important thing however is to listen for a while first and get the idea of what this contest wants you to do. The example I gave you is very generic and although is fairly typical, may not be correct for the contest you are working.

Also you need to make sure you get it right as the people you are trying to contact are in a contest to win and they are working hard, they have no time to deal with people who can't follow the rules and so if you mess up it is entirely possible that person will not call on you a second time.

If you want to try your hand at contesting check out https://www.contestcalendar.com/ and find one that interests you. I prefer open worldwide contests as they are easy and I get to try to communicate with people all over the world.

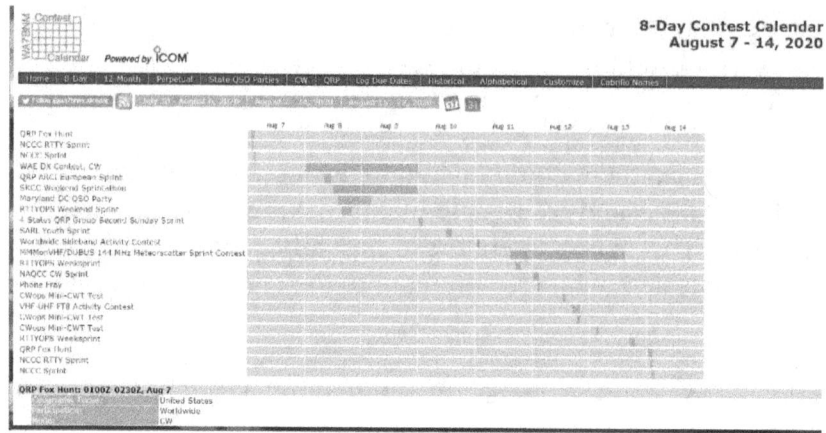

Some tips to getting called on is to never raise your voice, it tends to overmodulate and is considered rude. They are likely to ignore you until you go away. In fact, they may call you out and tell you to back off.

Try to change your inflection, monotones are easy to drown out in the pileup. Try Wd1r…..wdONER……..wD1r, etc. Changing things up makes you stand out without yelling or doing something stupid.

I have also heard people change the phonetics of their call with great success. Examples might include Whiskey Drinker 1 Round, Whiskey Delta 1 Radio, etc. If they can hear you they might find the approach novel and call on you although not everyone appreciates this approach.

Whatever you do, if he calls "ending in one romeo" and your call ends in something else, keep your mouth shut. Nothing makes everyone on the channel madder than to reply when it is obvious that he was not talking to you and you are simply trying to butt in to the quiet space. It is highly likely he will not call on you at all after you pull that.

So how do you find people to talk to in a contest? The best way is with a DX Cluster.

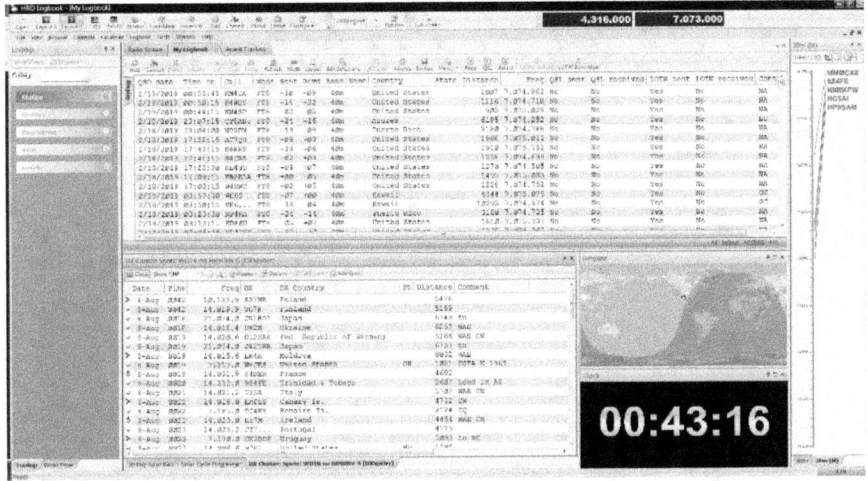

In the image above you see the chart type listing in the bottom center? That is the results of a DX Cluster and is being displayed in the logging portion of my Ham Radio Deluxe software.

This software is like a Swiss Army Knife for amateur radio operators as it can control your radio, manage your logging, display DX Cluster information, work with different digital modes, and much more all rolled into one.

The advantage here is that when I see someone I want to contact, I can double click the line and my radio will automatically change to the correct frequency to make the contact.

This is nice, but you don't have to buy software just to find potential contacts, use a website like https://www.dxwatch.com/

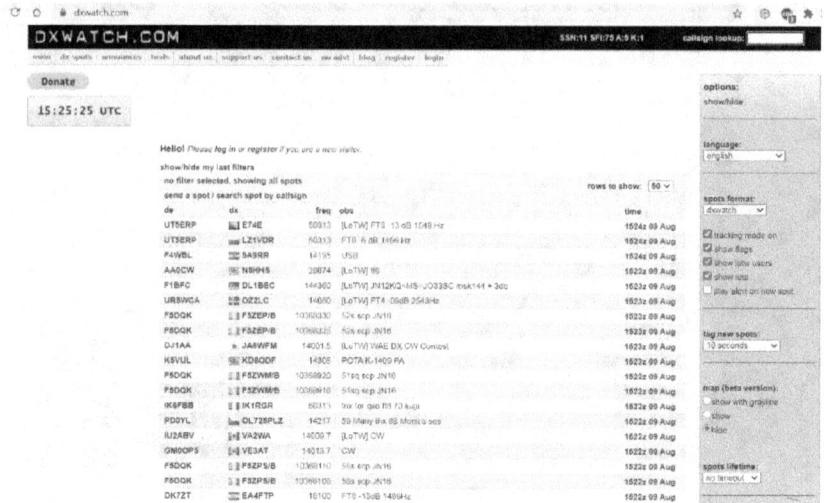

Here you can see much the same information but of course without the radio control features. Using the filters you can restrict what you see to specific bands, modes, country, zone, continent, and more.

It automatically updates so you can have this running on a computer while you are working contacts and keep an eye out for new contacts you want to try.

Two tips for you; always have spare paper and pencil around even when you are using a computer to keep your logging because things happen, and keep your computer time synchronized with and internet time source because accurate time matters in so many different ways when contesting.

On my Windows computer I use http://www.timesynctool.com/ as it is simple, free, and has proven to be reliable.

Skywarn

Skywarn is a group of volunteers trained and sponsored by the National Weather Service who provide reports of severe weather in their area to other radio operators and the National Weather Service.

These are not "storm chasers" like in the movies although they are more likely to be out in poor weather than regular sane people. If you see someone pulled over on the side of the road at the top of a hill, standing out in driving rain with a pair of binoculars and a radio, yeah, that's probably one of them.

Basically you attend a training class, which is actually a lot of fun. I have taken two over the years and both were quite entertaining as well as educational. Now most if not all the training is online so you don't even have to travel like I did, just head on over to https://www.weather.gov/oun/skywarn-spotter and get started!

You can use the training to join your local Skywarn network, or just to be more informed about severe weather.

I originally took the training classes to both understand severe weather better and to know how the local Skywarn nets worked so I knew where to listen in and what all I would hear. It really helped me be more prepared and have better real-time weather information in my area.

One book that really helped me, and that has been suggested by more than one Skywarn trainer I have met, is Ryan Henning's Field Guide to the Weather: Learn to Identify Clouds and Storms, Forecast the Weather, and Stay Safe. I really liked the sections on some of the common North American weather systems like Nor'easters, Texas Hookers (get your mind out of there), and Polar Vortexes.

While it is a small book, a little larger than this one, it is packed with amazing information and images. I like to keep it in my car for when I go storm spotting. I am sure the kindle version is good too, but I can't really see whipping out a kindle inside a pitch black car with a bunch of radio gear, maybe it's just me.

Local club events

I am not a big fan of local clubs for a lot of reasons. One of the biggest reasons is I don't care for sitting around a table and listen to a lot of people yack about things I am not into while eating doughnuts wasting my weekend mornings. I think I told you I am a little anti-social, right?

But that doesn't mean I don't enjoy occasionally participating in events. Local amateur radio clubs often participate in local events such as bike races and rides where they provide communications all along the route for things such as supply operations, rider monitoring, requests for emergency services, and just making sure the last rider has passed a rest area before closing that rest area.

There are lots of other places where volunteer radio operators are welcome as well. These events typically last a day or more and challenge your communications skills as well as provide a service that the people in the event really appreciate.

I have even had a bike rider stop and thank me, he was an amateur radio operator from out of town and wanted to express his appreciate for all the hard work that many never notice.

Challenges include setting up radio communications many miles from town in places where radio signals may have a difficult time reaching. This may require erecting antennas in trees, using your vehicle to hold the base of a free-standing antenna, or just little things like having enough battery power to operate all day.

Just contact your local club and ask them to let you know when any events come up you can help with. Of course they will try and get you to become a member and start showing up at their meetings, and that's OK too if that's your thing. Usually however they won't be that picky when the event comes up and they need volunteers and equipment ☺

If you do not know how to contact your local club you can always head on over to http://www.arrl.org/find-a-club and that should help you find it.

If your local club is not ARRL affiliated and you can't google them you might use the finder at the link above to find the nearest club to you, contact them and ask if they know of one that is closer to where you live. If not, consider helping the one you found, having to travel to get to their events is a great reason to excuse yourself from their regular club meetings.

In closing

I hope you have found some ideas to get you excited about amateur radio. These are some of the ideas that got me back into the hobby after I left for years because it had just become too social for me.

If I wanted social I would use a Facebook or Twitter account. I don't.

I worked hard to provide you with some good information, but I am human and probably made plenty of mistakes. If you spot one, hop on over to my website at besthamradio.com and use the contact form to drop me a line, call me an idiot, and give me a chance to correct the book so others can benefit from your wisdom.

Feel free to contact me if you have an addition too. If you found another aspect of the hobby that keeps you motivated while feeding your antisocial personality, I would love to hear about it and include it in the next version of this book.

If the book helps you, teaches you something, or is a good wedge for that table that used to rock back and forth, think about leaving a positive review on Amazon for the book as that really helps me more than you would think.

I wrote this during the great pandemic of 2020 so if you were one of those who got caught with your pants down and no toilet paper, it's OK, I understand.

73s!

More Information

Index

Amateur TV, 46
antenna, 9
APRS, 56
Arduino, 15
ARGO, 52
Automatic Packet Reporting, 56
Baofeng, 35
Bill Munsil, 46
Black Cat HF Weather Fax, 50
contest log, 64
contesting, 64
CubicSDR, 33
Davis Instruments, 58
digipeater, 57
Digital APRS, 57
dipole, 10
DX Cluster, 68
DXCC, 23
Earth-Moon-Earth, 40
events, 72
Fast Scan TV, 46
fox hunt, 34
FSTV, 46
FT8, 27
G5RV, 11
GPS, 56
HackRF, 33
Ham Radio Deluxe, 68
HF fax, 50
Igate, 58
Itty Bitty Radio Telescope, 27
James Stevens, 17
J-Poles, 10
JT65b, 42
Kite antennas, 48
local clubs, 72
Logbook of the World, 21
LotW, 21
Meteor scatter, 42
MMDVM, 14
Moon bounce, 40
Morse code, 52
National Weather Service, 70
Pacific Antenna, 37
QRP, 36
QRSS, 52
QSL cards, 22
Radio astronomy, 25
Raspberry Pi, 13
satellite, 38
SDR, 32
SignaLink, 29
Skywarn, 70
Slow scan TV, 44
Software Defined Radio, 32
SOTA, 54
SOTABeams, 49
Spectran, 53
SSTV, 44
Summits on the Air, 54
television, 46
Texas Hookers, 71
TNC, 57
Weak Signal Propagation Reporter, 60
weather stations, 58
Weatherfax, 50

WEFAX, 50
Worked All States, 23
WSJT-X, 28

WSPR, 60
yagi, 10

Glossary

AGC – Automatic Gain Control. A radio automatically manages the receiver's gain.

AM – Amplitude Modulation. Where the signal strength of the carrier wave is varied.

Amplifier – A set of circuits or a separate device used to increase the power output of a device such as a radio transmitter.

Anderson power poles – A connector type frequently used in amateur radio to connect devices to power supplies.

ANL – Automatic Noise Limiter. A circuit or device that helps to eliminate impulse and static noise peaks.

Antenna impedance – The impedance (resistance) of an antenna at the frequency at which it is designed to be used. Most antennas in amateur radio are designed to have a 50 ohm impedance.

Antenna matching – The act of making sure your antenna's impedance matches your transmitter output.

Antenna tuner – Circuit or device used to match an antenna impedance to that of your transmitter.

APC – Automatic Power Control. Automatically adjusts your transmitter's power output in case of an antenna that is not matched well.

APRS – Automatic Position Reporting System. A system that can report the position of the transmitter automatically at pre-determined intervals.

ARES – Amateur Radio Emergency Services. Part of the ARRL for communications during emergencies.

ARRL – American Radio Relay League. The national association responsible for many of the organizational duties of amateur radio in the United States including processing tests and licenses.

ASCII – American National Standard Code for Information Interchange. A set of 128 characters including upper and lower case letters, 0-9, and other characters typically found on a keyboard that form the basis of all modern computers and computer systems.

ATV – Amateur Television. Non-commercial television broadcast using SSTV, FSTV or digital methods by amateur radio operators.

Auto patch – A device that connects an amateur radio repeater with a telephone so that you can make phone calls using an amateur radio.

Backscatter – The result when radio waves are reflected back from the ionosphere.

Balun – A type of transformer that changes an unbalanced input into a balanced output.

Band – A defined range of radio frequencies.

Bandwidth – A measurement of the amount of radio frequency required for a particular type of transmission.

Bank – A group of memories in a radio.

Beacon – A device that emits a radio signal at intervals used to measure propagation characteristics.

BNC – Bayonet Neill-Concelman. One style of antenna connection that is no longer popular.

Busy channel lockout – A feature in radios that prohibit transmitting on a frequency that currently has traffic.

Call sign – An identifying set of letters and numbers for an amateur radio operator assigned by a country's radio administration.

Carrier – An unmodulated radio signal.

Clipping – The cutting off of the peaks of a transmitted signal caused by that signal being overdriven during amplification.

CTCSS – Continuous Tone Coded Squelch System. Widely used in repeaters it is a sub-audible tone that can be used to open squelch or allow access to the repeater.

CW – Carrier wave. Used to refer to Morse Code.

CW filter – A filter used with CW that improves reception by narrowing the bandwidth of the received signal.

D-STAR – Digital Smart Technologies for Amateur Radio. Primarily used by Kenwood and Icom, a digital radio protocol similar in function to DMR or Yaesu's Fusion.

DC – Direct current. The type of power usually associated with batteries and vehicles.

DCS – Digital Code Squelch. A digital version of CTCSS.

Deviation – The maximum amount a radio signal carries to either side of the carrier frequency.

Dipole – A bidirectional half-wave antenna.

Doppler shift – The shifting of frequencies as a transmitter approaches or recedes from the receiver. Typically noted in satellite and ISS communications.

Downconverter – A device that takes a high frequency signal and converts it into a lower frequency output signal.

Downlink – The transmission from a device such as a repeater to your radio.

DSP – Digital Signal Processor. An electronic circuit that digitally manipulates a signal with the intent of increasing the signal to noise ratio.

DTCS – Digital Tone Coded Squelch. A method to restrict communications between radios that have a specific digital signal transmitted.

DTMF – Dual Tone Multi-Frequency. The tones you hear when you press the numbers on a phone. Used to control repeaters, dial phones, and much more.

Dualwatch – The ability of a radio to monitor two frequencies at the same time.

Dummy load – A device connected in place of an antenna that does not transmit the signal from your transmitter but instead dissipates it for testing.

Duplexer – Used to split transmit and receive signals.

Duty cycle – The percentage of time that a transceiver is transmitting.

DX'pedition – An expedition whose purpose is to operate an amateur radio station, usually to very remote places such as uninhabited islands.

EME – Earth Moon Earth. A method of bouncing a signal off the moon and back to Earth for long distance communications. Also called moon bounce.

EMI – Electromagnetic Interference. Interference that can be heard on a radio that can be caused by lights, battery chargers, and other electrical devices.

Emission – The transmission of a signal.

F connector – A type of antenna connector found on some 70cm and 23cm antennas.

Fading – Where a signal seems to get stronger and weaker, often back and forth, and is caused by atmospheric conditions.

Feed point – The point at which a line is connected to the actual antenna.

Filter – A device or circuit designed to remove a portion of the signal, usually for the purpose of improving the signal to noise ratio.

FM – Frequency Modulation. Where the frequency of the carrier wave is varied.

FSTV – Fast Scan Tele Vision. What you think of as typical analog television.

Full duplex – A process where you can transmit and receive at the same time on two different frequencies. Think of a telephone as opposed to a CB radio.

Fuse – A device designed to fail under a specific load. Used to protect devices from a load that would damage that device.

Ground plane – An electrically conductive surface below and antenna used to couple the antenna to ground and serve as a reflecting surface for the radio waves.

Grounding – Connecting electrical connections to earth.

Ham – A licensed amateur radio operator.

Ham radio – Another term for amateur radio.

Harmonic – A multiple of a frequency.

Heat sink – A structure or device used to draw heat away from a device such as a transistor. These often have fins and can be made from aluminum or copper.

HF – High Frequency. Associated with frequencies between 3 and 30MHz.

HPF – High Pass Filter. Allows the higher portion of a signal to pass through while restricting the lower end of the signal.

Hz – Hertz. One oscillation of a signal.

IC – Integrated Circuit. A chip that combines multiple functions as if several electronic components were integrated into one.

Inverter – A device that converts DC voltage into AC voltage. Useful for running radios and other devices normally plugged into household electricity from batteries.

JT65 – A digital mode of communications involving a computer and radio that used to be very popular. Originally designed for extremely weak signal communications. Since replaced with FT8.

Knife edge – When a signal is refracted over tall objects such as buildings and mountain ranges.

LCD – Liquid Crystal Display. Commonly used in screens for radios, computer monitors, and televisions.

LED – Light Emitting Diode. Usually used for indicators, backlighting of displays, small light bulbs, and getting more common for computer monitors and televisions.

LF – Low Frequency. Generally refers to frequencies in the 30-300 kHz range.

Li-Ion – Lithium Ion. A rechargeable battery technology that is common with electronic devices today and has more capacity and less recharge memory than previous battery types.

LID – A radio operator who is rude and/or ignores current rules and operating procedures.

Logging software – Computer software used to record contacts made by amateur radio operators. Modern versions typically upload the contacts directly to online contact databases such as LotW maintained by the ARRL.

LPF – Low Pass Filter. Allows the lower portion of a signal to pass through while restricting the higher end of the signal.

LSB – Lower Side Band. A set of frequencies that are lower than the carrier frequency.

MARS – Aside from a planet, refers to Military Auxiliary Radio Service. An emergency communication response group made of volunteers and sponsored by the government.

Memory bank – see Bank.

Memory effect – The effect on a rechargeable battery that has been charged improperly which results in the battery having reduced capacity.

MIC – Short for microphone.

Mobile – Operating a radio at a non-fixed location such as in a car, on a bicycle, or walking.

Modulation – Changes a radio frequency carrier in order to add information to the signal.

NB – Noise Blanker. A circuit in a receiver that helps to reduce pulsing noises in a signal.

NiCd – Nickel Cadmium. An older type of rechargeable battery predating Ni-MH that suffered from extreme memory effect.

Ni-NH – Nickel-Metal Hydride. A battery technology prior to Li-Ion but after NiCd that improved on capacity, reduced memory effect, and reduced weight as compared to NiCd.

Notch filter – A circuit that rejects signals more sharply than a LPF or HPF.

Offset frequency – The difference between the transmit and receive frequencies, typically on a repeater.

PEP – Peak Envelope Power. The power transmitted at the maximum amplitude.

Power supply – A device used to convert from household current to the DC voltage needed to run a device. In amateur radio the DC voltage is usually 12V.

Priority watch – When a radio regularly checks a specific frequency for traffic and switches to that frequency when traffic is heard.

PTT – Push To Talk. The button on your radio or microphone used to begin transmission.

Reflected power – Power that is dissipated as heat instead of being converted to a transmitted signal, due to an improperly tuned antenna.

Refraction – When radio waves are bent back to earth off the ionosphere it is called refraction.

RF – Radio Frequency.

Repeater – A device that listens to a received signal and then retransmits that signal. Typically this is done on two different frequencies and in full duplex.

RF ground – Connecting equipment to earth ground to reduce interference.

RFI – Radio Frequency Interference.

RX – Short for receive.

S/N – Signal to Noise ratio.

Scan – The function of a radio where it changes frequencies or memories quickly looking for a transmitted signal.

Scan edge – The upper or lower frequency limit at which a radio can perform a scan.

Semi duplex – Where equipment such as a repeater will receive on one frequency and transmit on another, but not at the same time.

Sensitivity – The minimum signal a radio can detect.

Short Wave Radio – Refers to a radio that can receive signals from 10-100m or approximately 3-30MHz.

Simplex – A mode of operation that uses the same frequency for transmitting and receiving.

Skywarn – A group of volunteer amateur radio severe weather spotters trained by the National Weather Service.

SMA – Sub-Miniature A connector. A common type of antenna connection particularly on VHF/UHF handheld radios.

Split – When you operate by transmitting on one frequency and receiving on another.

SQL – Squelch. Allows you to quiet the radio until a signal of a particular strength is received.

SSB – Single SideBand. A set of frequencies that are either lower (LSB) or higher (USB) than the carrier frequency.

SSTV – Slow Scan Tele Vision. A method of transmitting photos or video frames over radio.

SWR – A ratio of the forward vs reflected power of an antenna. Generally the lower the SWR the better.

TNC – Terminal Node Controller. A type of modem used to transmit computer data over the radio which was once very popular.

TOT – Time Out Timer. Turns off the transmitter automatically after a certain length of time transmitting. Prevents overheating.

Transverter – A downconverter that works with both transmitting and receiving at the same time.

TSQL – Tone Squelch. A method to keep the squelch closed on a radio unless a specific sub-audible tone is received.

TX – Short for transmit.

UHF – Ultra High Frequency. Refers to signals in the 300MHz to 3GHz range.

UHF connector – A type of antenna connector also called a PL259 connection.

Uplink - The transmission to a device such as a repeater from your radio.

UTC – Universal Time Coordinated. Also called Greenwich Mean Time or GMT. A standard of time that does not change based on location or time of year.

VFO – Variable Frequency Oscillator. The mode on a radio where you can input any valid frequency for operation. Also is commonly used to refer to each "channel" or "bank" the radio can operate on. For example, dual band radios that display two frequencies on the screen are said to have two VFOs.

VHF – Very High Frequency. Refers to signals in the 30-300MHz range.

VOX – Voice Operated transmission) A feature that starts and stops transmission by listening for you to start and stop speaking.

Weather Alert – A station broadcasting weather information run by the National Oceanic and Atmospheric Administration.

Yagi – A type of highly directional antenna that looks similar to an old-fashioned TV antenna.

Q Codes

CODE	Description (informal or alternate)
QNA	Answer a net in order
QNF	A net is free
QNI	You can join a net
QNU	The net has traffic for you
QNX	You are released
QRA	My call sign/station name is (name)
QRG	The current frequency is (frequency)
QRH	Your frequency varies
QRI	Tone of transmission i.e. good, variable, bad
QRJ	Number of voice contacts
QRK	Readability of signal (intelligibility)
QRL	Are you busy?
QRM	Human-made interference
QRN	Natural interference
QRO	Increase power
QRP	Decrease power (low power)
QRQ	Send faster CW
QRS	Send slower CW
QRT	Stop operation
QRU	Have traffic for you
QRV	Ready
QRW	Inform xxx you are calling them on yyy
QRX	Standby until xxx
QRZ	Who is calling?
QSA	Signal strength
QSB	Fading signal
QSD	Keying defective
QSG	Send messages
QSH	Stay happy & healthy
QSK	Can you hear me between your signals (break in)
QSL	Acknowledge receipt (or QSL card)
QSM	Repeat last message
QSN	Do you hear me on xxx

QSO	Communication with (conversation, contact)
QSP	Relay message
QSR	Repeat call
QSS	Frequency to use
QST	Broadcast message
QSU	Send reply on xxx
QSW	Send on xxx
QSX	Listen on xxx
QSY	Switch to xxx (change frequency)
QSZ	Send word/group multiple times
QTA	Cancel message
QTC	Messages sent (or just message)
QTH	Position in long/lat (home)
QTR	The time
QTU	Times of operation
QTX	Listen for further communications until xxx
QUA	News of xxx
QUC	Last message received
QUD	Urgent message from xxx
QUE	Speak xxx language
QUF	Received distress signal from xxx

ITU Phonetic Alphabet

Alpha
Bravo
Charlie
Delta
Echo
Foxtrot
Golf
Hotel
India
Juliet
Kilo
Lima
Mike
November
Oscar
Papa
Quebec
Romeo
Sierra
Tango
Uniform
Victor
Whiskey
X-Ray
Yankee
Zulu

Other books by the author

The Baofeng UV-5R is probably the best selling amateur radio of all time, but it has one big problem, the manual. If you have a UV-5R and are struggling to get enough information out of the manual to actually use the radio then this book can help!

I will walk you through programming both simplex and repeater frequencies from the front keypad, and then show you how to install and use free software to make it even easier for you to program your radio. While the manual that comes with the radio tries to cover more information, what is presented here will get you up and running fast and easy.
https://amzn.to/3kyqCH5

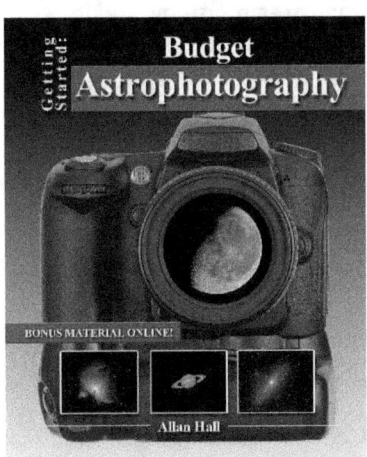

Allan Hall makes learning how to photograph the night sky easy with his new book Getting Started: Budget Astrophotography. In this guide, you will learn the fundamentals of astrophotography – what it is, how it's done, and how to do it yourself.

Getting Started: Budget Astrophotography is a great reference guide for beginners and amateur astrophotographers. If you have an interest in astronomy and want to capture what you've viewed through a telescope, doing so is possible from your own home. Hall's comprehensive guide also provides ideas about where to start (as in, what targets are best to photograph), where to find more information about astrophotography, and even a glossary of terms. Indulge your hobby and learn how to improve with Getting Started: Budget Astrophotography. https://amzn.to/3OF6CdF

Building, Setting Up, and Using the
MMDVM JumboSpot
for DMR, Fusion, & D-STAR

- Allan Hall -

https://amzn.to/31zg7u5

Do you want to get into digital amateur radio but are not sure where to start? Do you have a JumboSpot/MMDVM and not know how to set it up? Do you want to build your own JumboSpot?

If any of these are you, and you want to get going as fast as possible with no added fluff to sift through, then this book is for you!

From picking out a kit to build, through building it, downloading and installing the Pi-Star operating system, and configuring it to work with your radio, this book takes you step by step through everything you need to know to get up and running with DMR, Fusion, and D-STAR fast.

Setting Up a
Grandstream VOIP
Phone System

- Allan Hall -

Modern VOIP phone systems are amazing and the Grandstream systems are no exception. Where they often are lacking is their ease of setup or installation. In fact, out of the box they are often not much more than paperweights.

Many vendors who sell these systems want you to pay for programming called provisioning, and that can run hundreds or even thousands of dollars in addition to the initial price of the system. To make matters worse, the paperwork they want you to fill out can be overwhelming, simple systems with hundreds of questions. No thanks.

What you need is a simple blueprint that gets a basic system up and running so you can actually use your new VOIP system while you figure out what all fancy things you want it to do. That is what this book provides for you. https://amzn.to/3ab53ao

Notes

Notes

www.ingramcontent.com/pod-product-compliance
Lightning Source LLC
Chambersburg PA
CBHW071423210526
45465CB00001B/506